Assessing Revolutionary and Insurgent Strategies

INTRODUCTION TO ARIS

POCKET GUIDE

Johns Hopkins University Applied Physics
Laboratory (JHU/APL) Contributing Authors:

Meg Keiley-Listermann
Summer Newton
Melissa Ellison
W. Sam Lauber
Robert Leonhard
Jeff Macris
Christina Pikas

United States Army Special Operations Command

Printed in the USA by the Government Printing Office

Cite me as:

Keiley-Listermann, Meg, et al. *Introduction to ARIS: Pocket Guide*. Laurel, MD: JHU/APL, 2019.

ASSESSING REVOLUTIONARY AND INSURGENT STRATEGIES

The Assessing Revolutionary and Insurgent Strategies (ARIS) series consists of a set of case studies and research volumes conducted for the US Army Special Operations Command by the National Security Analysis Department of the Johns Hopkins University Applied Physics Laboratory. The purpose of the ARIS series is to produce a collection of academically rigorous yet operationally relevant research materials to develop and illustrate a common understanding of insurgency and revolution. This research, intended to form a bedrock body of knowledge for members of the Special Forces, will allow users to distill vast amounts of material from a wide array of campaigns and extract relevant lessons, thereby enabling the development of future doctrine, professional education, and training.

The ARIS series follows in the tradition of research conducted by the Special Operations Research Office (SORO) of American University in the 1950s and 1960s, by adding new research to that body of work and in several instances releasing updated editions of original SORO studies.

RECENT VOLUMES IN THE ARIS SERIES

Casebook on Insurgency and Revolutionary Warfare: Volume I and Volume II

Human Factors Considerations of Undergrounds in Insurgencies (2013)

Undergrounds in Insurgent, Revolutionary, and Resistance Warfare (2013)

Understanding States of Resistance (2019)

Legal Implications of the Status of Persons in Resistance (2015)

Threshold of Violence (2019)

"Little Green Men": A Primer on Modern Russian Unconventional Warfare, Ukraine 2013-2014 (2015)

TABLE OF CONTENTS

LIST OF ILLUSTRATIONS

INTRODUCTION

The Assessing Revolutionary and Insurgent Strategies (ARIS) project consists of research on the phenomenon of resistance conducted by the Johns Hopkins University Applied Physics Laboratory (JHU/APL) for the US Army Special Operations Command (USASOC) G3X Special Programs Division.

The purpose of ARIS is to produce a collection of academically rigorous *and* operationally relevant research materials that addresses challenges that the US Army Special Operations Forces (ARSOF) encounter when conducting their mission sets. This multidisciplinary research seeks to inform the ARSOF soldier's mission preparation through applied learning of historical case studies, topically focused studies, and selected specialized topics across the global security landscape.

The ARIS project further seeks to expand the resistance studies by synthesizing academic research, informing doctrinal development and professional education and training, and translating academic findings into operational impact. ARIS builds on the work conducted by American University's Special Operations Research Office (SORO).

This *Introduction to ARIS* pocket guide serves two purposes: (1) to familiarize the soldier with an overview into the scientific study of resistance and (2) to provide the ARSOF soldier with an academic synopsis of ARIS publications.

BACKGROUND

In the Cold War, the US Army funded a series of academic studies of revolutionary and insurgent warfare topics. In the 1950s, the US Army turned to American University in Washington, DC to craft study materials connected to nations central to the United States' ongoing political and military struggle against the Soviet Union and its allies. Those academic materials included case studies, topical investigations, and bibliographies. American University's SORO in 1959, for example, published *Area Handbook for Hungary* and *Psychological Operations: Egypt*, as well as *An Introduction to Wartime* leaflets.

In the 1960s, the SORO published scores of additional texts, including case studies on areas such as Vietnam, Cuba, and Algeria. It also published topical investigations of interest to US Special Forces, such as *Working with Peoples in Developing Areas: One Task of the American Soldier Overseas* (1966), *Human Factors of Undergrounds in Insurgencies*, as well as a bibliography devoted to counterinsurgency. In total, SORO published over 150 texts for the US Army Special Forces. The number of SORO publications decreased substantially by 1966 and ceased in 1968 when American University discontinued the program.

REIMAGINING SORO AS ARIS

ARSOF operators should understand resistance as the object of their profession, according to current director of the USASOC G3-X, Mr. Paul Tompkins. The utility of academic studies on the phenomenon of resistance was embraced by Mr. Tompkins early on in his career. In 1979, Mr. Tompkins attended the Special Operations Qualification course and was given a copy of the SORO publication, *Human Factors Considerations of Undergrounds in Insurgencies*. This book, and the later SORO publication *Undergrounds in Insurgent, Revolutionary, and Resistance Warfare*, prepared him for the work he would undertake in the 1980s as a member of the 10th Special Forces Airborne Division, planning sensitive stay-behind operations in Eastern Germany and infiltration operations to support resistance activities in Poland and the Soviet Union.

Mr. Tompkins additionally integrated his academic knowledge of resistance into operational planning executed in both the training of ARSOF soldiers and their deployments. During the 1990s, Mr. Tompkins used the SORO publications as a planner and as an instructor to deepen his understanding of the phenomenon of resistance and to share that knowledge with his students. When he served as the team lead for Operation Detachment Alpha (ODA) 26, he planned resistance operations for the Kurds in support of Desert Storm and Operation Provide Hope. As the plans officer for 1st Battalion 10th Special Forces Group (A), he planned seven crisis operations in Africa.

As a commander and chief instructor at the US Army John F. Kennedy Special Warfare Center and School (SWCS), Mr. Tompkins once again used these foundational studies on resistance to shape the ARSOF soldiers' understanding and application of their mission. Paul Tompkins additionally served as the principal author of the classified Field Manual 3- 05.2 20, "Special Forces Advanced Special Operations," and was integral to the development of unconventional warfare (UW) plans that directly impacted operations in Afghanistan and the Global War on Terror.

In 2004, he became the deputy chief of the USASOC G3X Special Programs Division. Mr. Tompkins reflected on those previous SORO publications and how they informed his career as a highly decorated, accomplished soldier. He began to refresh the publications with new research and contemporary insights, and in 2009, he became the chief of USASOC G3X. Understanding the importance of applying multi-disciplinary research to the ARSOF soldiers' kit, Mr. Tompkins set out to revitalize the study of the phenomenon of resistance by chartering the ARIS program as a collaborative project with JHU/APL in 2009.

JHU/APL and the ARIS Program

Similar to the studies published by American University in the 1950s and 1960s, JHU/APL's ARIS works include a mix of case studies, topical investigations, and bibliographies—all aimed to provide special warfare soldiers and their instructors with easy-to-digest, academically sound learning materials. To that end, one of JHU/APL's first ARIS publications was the "Irregular Warfare Annotated Bibliography" (2011), aimed largely at instructors to assist them in locating current scholarship on revolutionary and insurgent strategies.

Shortly after, ARIS published two collections of case studies that included the 1979 Iranian Revolution, the Provisionary Irish Republican Army (IRA), and the Fuerzas Armadas Revolucionarias de Colombia (FARC) insurgency in Colombia. The case study on FARC, for example, traces the development of the group from its origins in 1966 to the present. The study gives soldiers a historical grounding in this well-known case of a long-running insurgency. The case study details how the FARC won allegiance from supporters, examines the tactics it employed during some of the most notorious military campaigns in 1998, and identifies the challenges that the FARC encountered in 1999 as a result of Plan Colombia (an effort to strengthen the Columbian government, integrate isolated areas, and enforce current laws). ARSOF operators can draw upon the historical lessons of this case study (and others) to inform their planning, operational activities, and professional development.

Movement Toward Topical Studies

Around 2015, the ARIS project undertook topical studies on subjects of importance to the ARSOF soldier. ARIS work featured new, original research books, fresh topics in book chapters enclosed in existing publications, and articles published in *Special Warfare* that explore resistance. An example is *Special Topics in Irregular Warfare: Understanding Resistance,* where authors explore subjects like the phases of resistance, how resistance groups generate legitimacy, and the use of information technology by resistance movements.

The ARIS project also expanded to include an online portal where civilian and military instructors can access ARIS materials to support their missions. In the fall of 2019, this content was transferred to the USASOC website where ARIS publications can be accessed and downloaded (scan the QR code).

Striving to deepen the application of academic study to operational requirements, the ARIS project began to field requests on contemporary challenges facing the ARSOF soldier. Three examples of the special topic studies are *Little Green Men: A Primer on Modern Russian Unconventional Warfare in the Ukraine; Legal Implications of the Status of Persons in Resistance*; and *Narratives and Competing Messages.*

Little Green Men offers the soldier a theory of Russian UW, tracing its development over decades in Russian activities in Lithuania, Chechnya, and Georgia as it reached its zenith in Russia's invasion of Ukraine (2013-2014). This study provides the soldier with a detailed synthesis of the tactics, techniques, and procedures (TTPs) that the Russians used, the hallmarks of which are deception, psychological manipulation, and domination of the information domain. *Little Green Men* gives the Special Operations community an understanding, grounded in contemporary and historical events, of one of the United States' rising foes (Russia).

Legal Implications of the Status of Persons in Resistance examines the legal status of resistance movements and US military members conducting UW in support of foreign resistance movements. Because the legal status of the resistance movement is directly tied to the nature of its activities, the status may impact how the US military advises and supports the group, and it may impact the service members' status as well. The legal status of resistance and military members can vary from country to country and over time within a single country. To address the challenge caused by confusion over the legal status, the authors take a novel approach, creating a continuum of a resistance movement's activities, from nonviolent to violent. This continuum aids soldiers in determining the resistance movement's methods and identifying the movement's international recognition and legal status. Through case studies and operational vignettes, this study makes the legal issues and nuances accessible for the soldier and visualizes the interplay between a movement's activities/international recognition and US policymakers' interpretations of law. This study helps soldiers to understand and anticipate the types of contexts in which they will be employed and to obtain a general understanding of the relevance of legal status for themselves and those they advise.

Scientific Study of Resistance

Understanding the usefulness of these special topical studies, the sponsor recently asked JHU/APL to pursue another iteration of research—building a science of resistance. This type of effort requires mining the academic literature for studies that address what the Special Operations community refers to as "resistance." This effort brought the political science and sociology fields of work on civil war, insurgency, social movements, and contentious politics to bear. These studies offer hypotheses, examine theories, and explore variables that shape the outcome of resistance.

Working toward a science of resistance also requires the examination of the universe of relevant historical cases. While individual case studies foster an understanding of that case and in developing hypotheses, case study findings are only applicable to the case upon which they developed. In contrast, developing a science of resistance requires that findings are developed by assessing all cases or a representative sample of those cases. Therefore, findings from such research are generalizable to the many environments in which the ARSOF soldier might find themselves.

A recent effort toward developing a science of resistance is *The Day After Overthrow: How the Behavior of the State and the Resistance Shapes Post-Overthrow Outcomes*. This forthcoming study seeks to determine the characteristics of a resistance movement that influence the nature of post-conflict governance. The study identifies post World War II cases of state overthrow from a dataset compiled by the Center for Systemic Peace, conceptualizes a novel magnitude of overthrow scale, and selects a portion of the scale for analysis. The analysis employs a statistical social science methodology to identify factors correlated with particular post-conflict governance outcomes. Future research will expand this study by examining cases along the entire magnitude of overthrow scale and conducting a statistical analysis that yields causal findings about the factors that determine the nature of post-conflict governance.

Emphasizing ARIS Usefulness and Effectiveness for the Soldier

The primary goal of the ARIS work has been to consistently provide operationally relevant and useful research and materials to the ARSOF soldier. It is only when soldiers turn to the ARIS work as an essential resource as they prepare for and execute mission sets in a deployment that ARIS will be a true success. Understanding what will be useful to the soldier is an important task for ARIS and requires a research effort of its own. In the spring and summer of 2018, a research team from JHU/APL deployed to gather insights into the capabilities that ARSOF soldiers want and need to achieve efficient and effective products in consumable, digestible, and applicable forms.

The future of ARIS must ensure that the research ARIS conducts and the material it develops will be useful to, and used by, the soldier. It is important to note that what will be useful to the soldier is constantly evolving; accordingly, satisfying soldiers' needs will be an iterative process based on an open dialogue with them on user requirements and access to cutting-edge academic and technological research to develop relevant and rigorous content to enhance operational effectiveness in the execution of specific mission sets.

In turn, the future of ARIS must be informed by the scientific study of resistance. The analytical approach must be repeatable, reliable, explainable, and traceable. Studies must continue to leverage exacting methodologies and cutting-edge research from political science, sociology, anthropology, psychology, history, economics, and law, but studies should consider emerging fields as well.

PART I: RESISTANCE STUDIES –
WHAT IS RESISTANCE?

"It is not to political leaders our people must look, but to themselves. Leaders are but individuals, and individuals are imperfect, liable to error and weakness. The strength of the nation will be the strength of the spirit of the whole people."

—*Michael Collins[1]*

According to Joint Publication 3-05, "Special Operations," a resistance movement is:

An organized effort by some portion of the civil population of a country to resist the legally established government or an occupying power and to disrupt civil order and stability.[2]

When encountering resistance movements, a number of questions will arise for the soldier, such as: Who is in this movement? What are the characteristics of the leadership, the organizational structure of the movement, and the recruitment strategy of the movement? What are the characteristics of the government the movement resists? Why do they resist? What are the movement's objectives and strategy to achieve those objectives? What tactics does the movement use? What resources, logistics, and equipment do they require? How is the movement supported (or not) by third-party actors?

Exploring these questions reveals the broadness of the category of resistance movements. While each resistance movement comprises people who seek to disrupt the established government or occupying power's civil order and stability, its motives, tactics, and external support can widely vary. The resistance movement can be motivated by retribution for the government's actions, ethnic inequality, the desire for power, or socioeconomic inequality, among others. Regarding tactics, some resistance movements are nonviolent, while others are violent. Even among resistance movements that employ violent tactics, there can be a wide variation in the types of violent tactics (e.g., improvised explosive devices, hijackings, kidnapping, and murder), when those tactics are employed in the life cycle of the resistance movement, and the extent to which those tactics are employed (e.g., to shut down ports, intimidate voters in an election, influence government policy). Finally, resistance movements can differ with regard to external support: Does the movement receive external support? Who is the external support from (state or nonstate actors)? What form does the external support take (cash, arms, training, mercenaries)?

This brief discussion demonstrates the distinctions between one resistance movement and another. These differences are all consequential to the soldier, as it is imperative that the ARSOF soldier can assess the causes (motivations) and environment (resources, organizational relationships, etc.) that shaped, constrained, and enabled the actors in the resistance movement it encounters. Therefore, a primary imperative of the ARIS body of work is to help the soldier understand resistance movements so that s/he can make informed choices in the field to bring about desired outcomes.

In understanding this multifaceted and nuanced phenomena of resistance, it is important to note that resistance has been well studied in the academic disciplines of political science, history, sociology, and anthropology, among others. Soldiers may not be familiar with the breadth of this work because it is vast, and these fields do not always use the term *resistance*. In some of these academic fields, soldiers use the term resistance when they also refer to conflict, civil war, social movements, revolution, rebellion, insurgency, civil disobedience, and protest. ARIS seeks to leverage this literature to provide the soldier with a robust understanding of the constructs encompassed by resistance.

To understand the breadth of research pertinent to resistance, ARIS employed a bibliometric methodology for this pocket guide. Bibliometric analysis involves a quantitative study of publications for empirical data analysis. Bibliometric analysis uses subject and topic clusters, citation trajectories, reference year spectroscopy, and algorithmic historiography to explore the fields where scholars publish on resistance-based topics and the significant articles within those publications in the field of resistance studies.

To begin, a keyword search of published academic literature using terms such as resistance, conflict, violence, strife, civil disobedience, ethnic conflict, insurrection, rebellion, or coup identified almost twelve thousand articles, chapters, and books. To narrow such a vast amount of data, the analysis effort first focused on the publications that held the greatest significance. Significance, in this context, means studies that furthered one's understanding of the topic, changed the soldier's views, and served as a foundation for subsequent research. The metrics of significance detailed in this *Introduction to ARIS* are frequency of resistance keywords in academic journals, frequency of resistance case studies in academic journals, and academic research cited by other academics.

Across all areas of scholarly study, articles on resistance appear mostly in the fields of political science, economics, and sociology, but also in criminal justice, public health, philosophy, and psychology. Figure 1 shows an overlay of journals containing articles on resistance that are found within a topic-based map of scholarly journals. The figure depicts clusters of resistance-based articles in academic journals, such as the *Journal of Peace Research*, *Democratization*, *Journal of East African Studies*, *Geoforum*, *Journal of Economic History*,

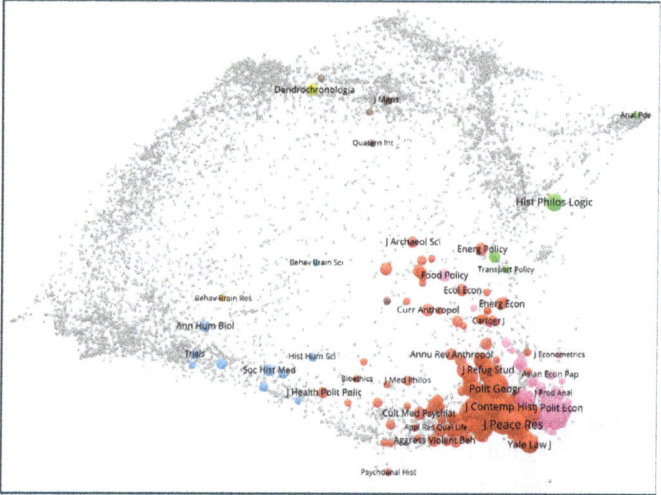

Figure 1. Subject cluster map of scholarly journals overlaid with journals containing resistance studies.

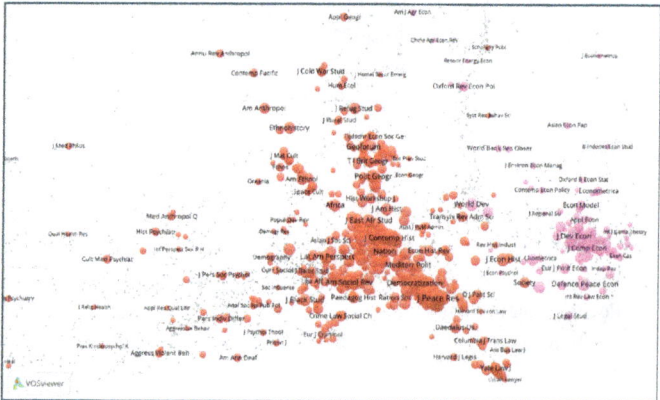

Figure 2. Highest frequency cluster of resistance study journals.

and *Latin American Studies*. Figure 2 provides a closer examination of the journals that contain a high number of resistance studies.

To further examine the body of significant resistance studies, one should query the most frequently studied topics. When scholars write about resistance, they support their arguments by citing or referring to primary sources that include original documents, interviews with participants, and artifacts, as well as secondary sources. Patterns identified in the citations of these various publications demonstrate a trend of time periods of the greatest interest to subsequent researchers, the most significant works in the field of resistance studies, as well as the locations of case studies cited frequently by researchers.

More than eleven thousand articles were grouped into topics using a machine-learning algorithm. In Figure 3, each circle represents an article. The colors show topic clusters of articles. The more related the content, the tighter the topic cluster. The topics with the most resistance-based topics are causes of conflict, ethnic conflict, the English civil war, democratization and regime change, rebels and civil wars, Russia/Soviet/post-Soviet interactions, Middle East North Africa region, and methods and models of mobilization. For each article, the most frequently mentioned country was extracted. Figure 4 shows the countries using the colors from the topic clusters in Figure 3. The most prolific studies are on the American Civil War, the Spanish Civil War, Uganda's civil upheavals, the Chinese Revolution, and Sri Lanka's Civil War. It should also be noted that the various Scottish and Irish uprisings in the eighteenth century have been well studied as well.

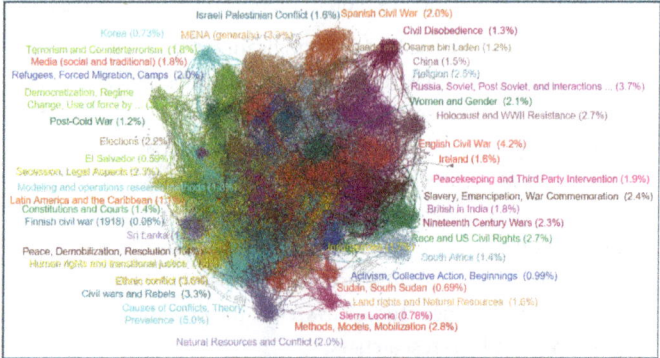

Figure 3. Topic clusters in studies of resistance.

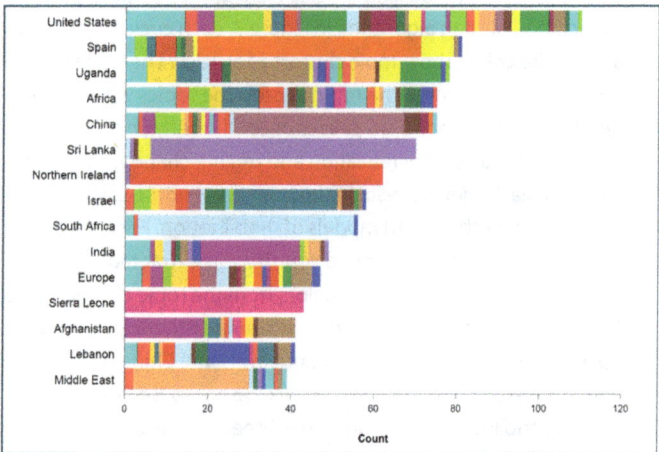

Colors represent the topic clusters shown in Figure 3. Color coding is detailed in the appendix.

Figure 4. Countries mentioned most frequently in studies of resistance.

What does the frequency of case studies mean for the ARSOF soldier interested in resistance studies? Perhaps the soldier's area of operations is not well researched. However, the variables involved in a resistance—the actors, causes, organization, actions, and environment[3]—may in fact be rigorously examined in these studies. This is an important hallmark of the ARIS studies—providing quality research for the soldier's consumption to inform, educate, and prepare the soldier for his/her deployment. Further refinement of the truly significant studies will only enhance this impact.

Figure 5 shows a geographical view of the subjects of resistance studies. The white lines in the figure show modern political boundaries, but resistance can be within a state or across borders. Multiple nodes in a country depict different levels of granularity in coverage.

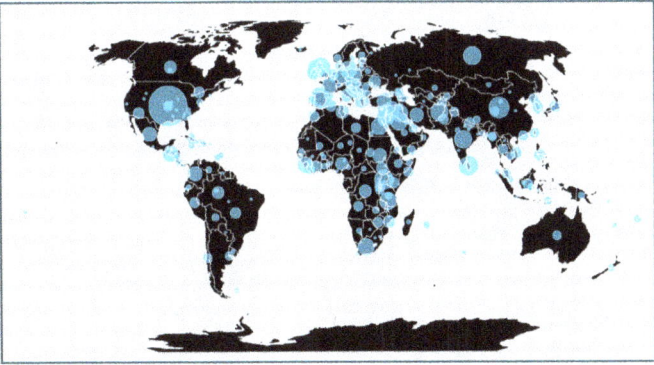

Modern political borders are shown in white. The size of the circle corresponds to the number of articles.

Figure 5. Depiction of the geographical subjects of resistance studies.

Finally, to further define metrics that identify which studies are significant pieces in the field of resistance studies, one should examine the work most cited by other academics. Figure 6 depicts a bibliometric graph of the authors cited by other academics in their research on resistance. Each circle represents an important, well-cited article and is labeled with the author's last name. Calling attention to the circle at the top of the figure, important studies of note from the 1960s and 1970s include works by Mancur Olson, Samuel Huntington, Ted Robert Gurr, Arendt Lijphart, Charles Tilly, and Theda Skocpol, which are foundational to recent scholarship. Their work influenced the likes of Benedict Anderson, Kate Fearon, and Paul Collier in the 1990s and Bruce Bueno de Mesquita and Haivard Hegre in the 2000s.

Therefore, one should consider the impact of a study, the lessons from within that can be applied elsewhere, as a feature of its significance. Some of the most significant articles sought to provide an overarching explanation for a key concept. For example, Ted Robert Gurr's *Why Men Rebel* uses a social psychological approach to explain political violence.[4] It won several awards when first published and was re-released forty years later. Gurr's work shaped a generation of research, but it also received criticism for not emphasizing societal level factors described in other work.[5] Likewise, Samuel Huntington's *Clash of Civilizations*, is controversial because it predicts that religious and cultural divisions will be paramount in future resistance movements.[6] It should be noted that critics of Huntington state that this view is overly simplistic and only serves to justify Western military interventions and false divisions between Western and other world cultures. Despite such criticism, his research continues to serve as foundational literature in the study of resistance.

Conversely, some significant articles provide reviews of bodies of work or periods of time that become foundational because it covers a wide scope of topics by subject or time period. For example, James D. Fearon and David D. Laitin authored a series of reviews on ethnicity and civil war.[7] Haivard Hegre and his colleagues reviewed civil wars from 1816 to 1992 to discover overarching commonalities and to then, in a later study, make predictions.[8]

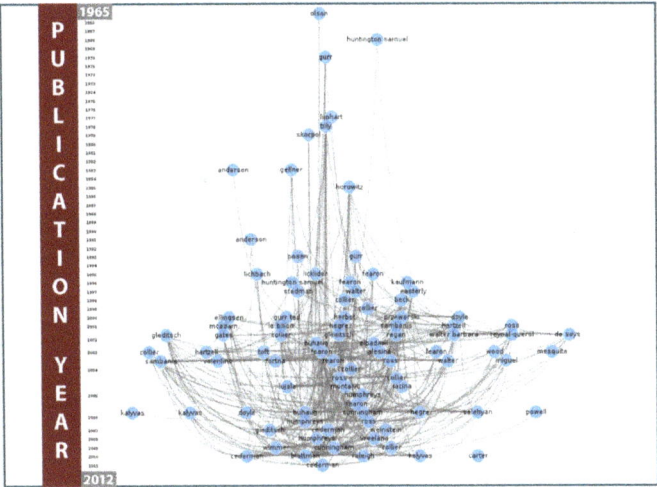

Figure 6. Algorithmic historiography citations.

To examine the influence of the academic research, one should also explore the institutional affiliation of the researcher. Five institutions were frequently cited in bibliometric queries of resistance terminology. Table 1 captures these institutions, the authors affiliated with them, and the keyword topic of research. These authors represent the cutting-edge research on resistance studies and the foundational research in a discipline, according to the frequency that their peers cite their work. By following the scholars that cite this research, one can continue to identify significant advances in future resistance studies.

In conclusion, part one of this *Introduction to ARIS* pocket guide provided a foundation to the scientific study of resistance. Reviewing bibliometric analysis, the ARSOF soldier can identify the significant case studies, topics, researchers, and citations in resistance studies. The following section will focus on the body of ARIS work that contributed to the advancement of the academic pursuit of resistance studies.

Table 1. Paradigm-shaping scholars and institutions.

Organization	Authors	Keywords
Columbia University (New York, NY)	Humphreys, Macartan Blattman, Christopher Foner, Eric	Civil war Greed Fractionalization Linguistic diversity
Harvard University (Cambridge, MA)	Baum, Matthew A. Toft, Monica D. Cohen, Dara K. Zhukov, Yuri M. Faust, Drew G.	Violence Third-party intervention Caucasus Exceptionalism Partition
Yale University (New Haven, CT)	Kalyvas, Stathis N. Sambanis, Nicholas Lyall, Jason Kocher, Matthew	Reintegration West Africa Armed parties Micro-level studies

Organization	Authors	Keywords
Oxford University (Oxford, UK)	Collier, Paul	International relations
	Hoeffler, Anke	Political ideology
	Buchanan, Tom	Civilian agency
	Soderbom, Mans	Fieldwork in conflict zones
Uppsala University (Uppsala, Sweden)	Hultman, Lisa	Violent conflict
	Fjelde, Hanne	Military intervention
Peace Research Institute of Oslo (PRIO), Oslo, Norway	Gleditsch, Kristian S.	Civil war onset
	Hegre, Haivard	Transitional justice
	Gleditsch, Nils P.	Urban
	Buhaug, Halvard	Conflict termination

PART II: ARIS PUBLICATIONS

"You do not become a 'dissident' just because you decide one day to take up this most unusual career. You are thrown into it by your personal sense of responsibility, combined with a complex set of external circumstances. You are cast out of the existing structures and placed in a position of conflict with them. It begins as an attempt to do your work well, and ends with being branded an enemy of society."

—*Václav Havel*[9]

ARIS PUBLICATIONS AND ICONS

For the ARSOF soldier to understand the overall topic and intent of ARIS publications, this pocket guide provides small abstracts that outline the basic structure of each publication since 2009. Figure 7 illustrates the timeline of the ARIS studies in both a chronological and geographical view. The ARIS program utilizes seven academic disciplines to study the phenomenon of resistance to inform the ARSOF soldier for operational planning and field operations (see Figure 8). The goal of the ARIS program is to produce academically rigorous and operationally relevant research used by and useful to the ARSOF soldier and to foster the persistent study of resistance. The publications are categorized by military operational specialty (MOS) content and the Theater Special Operation Command (TSOC) area of responsibility (AOR) of the resistance movement discussed in that publication. Icons representing the MOSs are placed by each publication's abstract. Tables identifying the TSOC AORs are within the text of the abstract, if applicable. Figure 9 depicts the icon legend for MOSs. If a publication contains content relevant to all ARSOF, the Special Forces icon is used.

Non-Conventional Assisted Recovery (NAR)

2009
USASOC G3X charters the ARIS project to produce operationally relevant case studies on resistance.

2010
Irregular Warfare Annotated Bibliography

2011
Casebook on Insurgency and Revolutionary Warfare Volume II: (1962-2009)

2012
Undergrounds in Insurgent, Revolutionary, and Resistance Warfare

Human Factors Considerations of Undergrounds in Insurgencies

Casebook on Insurgency and Revolutionary Warfare Volume I: (1933-1962)

2013
Case Studies in Insurgency and Revolutionary Warfare: Colombia (1964-2009)

2014
Case Studies in Insurgency and Revolutionary Warfare – Sri Lanka (1976-2009)

Legal Implications of the Status of Persons in resistance

2015
"Little Green Men": A primer on Modern Russian Unconventional Warfare, Ukraine (2013-2014)

Case Studies in Insurgency and Revolutionary Warfare: Palestine Series Volume I - The Zionist Insurgency (1890-1960)

Special Topics in Irregular Warfare: Understanding Resistance

2016
UW Pocket Guide Volume I

2017
Unconventional Warfare Case Study: The Relationship Between Iran and Lebanese Hizbollah

Unconventional Warfare Study Research and Writing Guide

Insurgency Study and Writing Guide

Narratives and Competing Messages

Understanding States of Resistance

The Science of Resistance

2018
Case Studies in Insurgency and Revolutionary Warfare: Palestine Series Volume II The Palestinian Arab Insurgency (1890-2010)

ARIS Codebook

Threshold of Violence

Resistance in the Cyber Domain

2019
Unconventional Warfare Case Study: The Rhodesian Insurgency and the Role of External Support (1961-1979)

The Day after Overthrow: How the Behavior of the State and the Resistance Shapes Post-Overthrow Outcomes

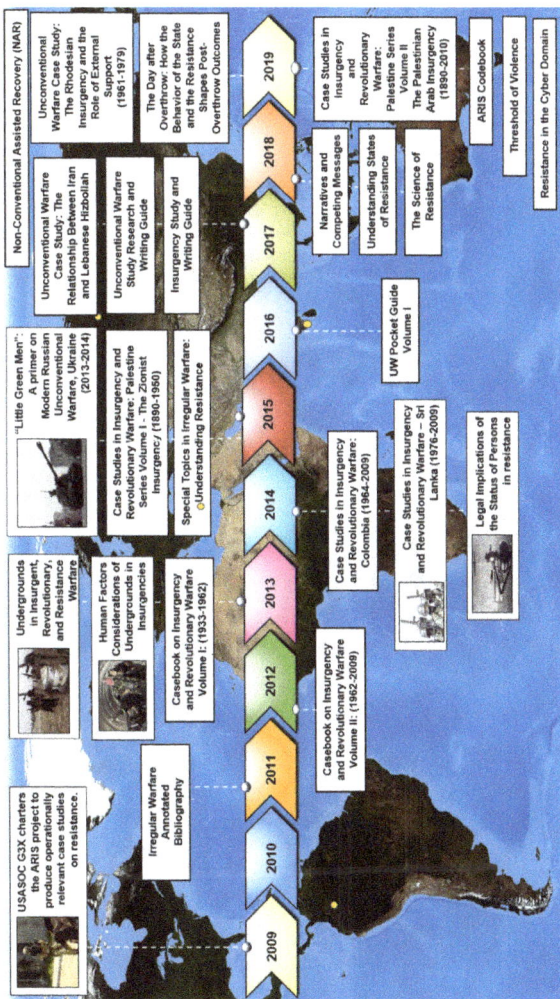

Figure 7. ARIS publications, 2009-2019.

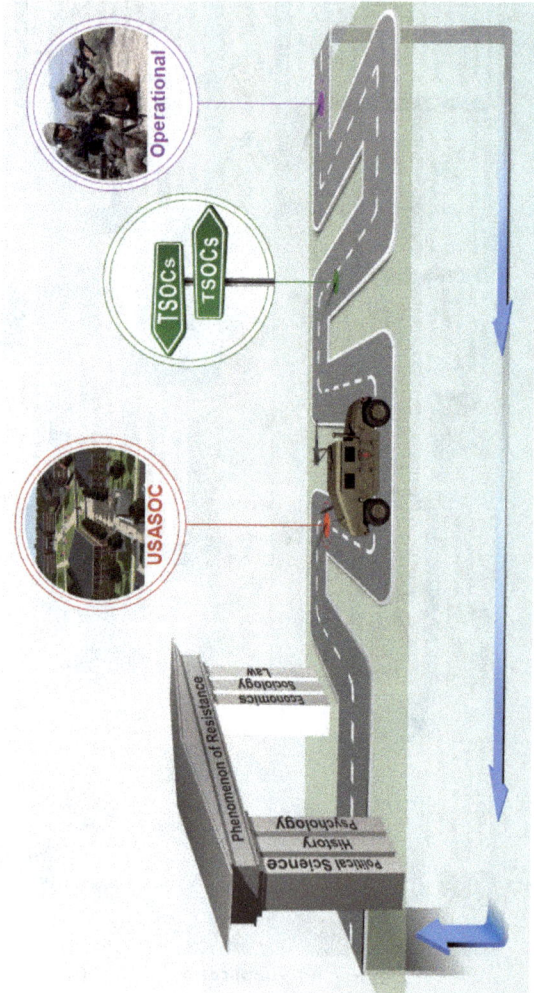

Figure 8. ARIS icon legend for MOS.

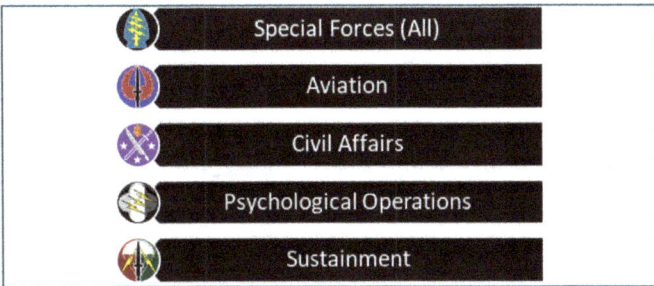

Figure 9. ARSOF iconography.

Irregular Warfare Annotated Bibliography

This volume is an annotated bibliography of sources, spanning military history and theory, political science, sociology, psychology, and anthropology related to the practice and study of irregular warfare. Sources encompass material relevant to Special Operations Forces' missions, including foreign internal defense, counterterrorism, unconventional warfare, and counterinsurgency. Each annotation is accompanied by a link that directs soldiers to a library database, WorldCat, where the source may be located. Introductory material discusses the concept of irregular warfare and the Special Operations Forces' missions associated with it. The soldier can use this bibliography as a starting point for future, independent research.

2012

Casebook on Insurgency and Revolutionary Warfare Volume II: 1962-2009

This volume contains case studies of twenty-three resistance movements.

SOCAFRICA	Hutu-Tutsi Genocides
	Egyptian Islamic Jihad (EIJ)
	Movement for the Emancipation of the Niger Delta (MEND)
	Revolutionary United Front (RUF) – Sierra Leone
SOCCENT	1979 Iranian Revolution
	Palestine Liberation
	Afghan Mujahidin: 1979-1989
	Hizbollah: 1982-2009
	Hizbul Mujahideen Taliban: 1994-2009
	Akl Qaeda: 1988-2001
SOCEUR	Kosovo Liberation Army (KLA): 1996-1999
	Provisional Irish Republican Army (PIRA): 1969-2001
	Chechen Revolution: 1991-2002
	Orange Revolution of Ukraine: 2004-2005
	Solidarity

SOCPAC	New People's Army (NPA)
	Karen National Liberation Army (KNLA)
	Liberation Tigers of Tamil Eelam (LTTE)
	Viet Cong: 1954-1976

SOCSOUTH	Fuerzas Armadas Revolucionarias de Colombia (FARC)
	Sendero Luminoso (Shining Path)
	Frente Farabundo Marti Para la Liberacion Nacional (FMLN)

The cases included in the work, dating from 1962 to 2009, were selected to cover a wide range of outcomes, geographical regions, historical periods, and motivations. The cases are grouped into five sections: revolutions seeking a significantly modified government, cases where identity or ethnicity underpins the movement, cases seeking to drive out a foreign power, revolutions associated with religious fundamentalism, and movements motivated by modernization reforms. The cases selected represent a spectrum of violent and nonviolent campaigns. Each case study is organized along a research framework with a case synopsis, an environmental assessment, the form and characteristics of the resistance movement, and an analysis of the short- and long-term effects of the conflict.

2013

Casebook on Insurgency and Revolutionary Warfare Volume I: 1933-1962

This volume contains case studies of twenty-three resistance movements.

SOCAFRICA	Tunisia: 1950-1954 Algeria: 1954-1962 French Cameroun: 1956-1960 Congo: 1960 Sudan: 1958
SOCCENT	Iraq*: 1936 and 1958 Egypt: 1952 Iran: 1953
SOCEUR	Germany: 1933 Spain: 1936 Hungary: 1956 Czechoslovakia: 1948
SOCPAC	Vietnam: 1946-1954 Indonesia: 1945-1949 Malaya: 1948-1957 Korea: 1960 China: 1927-1949

SOCSOUTH

Guatemala: 1944

Venezuela: 1945

Argentina: 1943

Bolivia: 1952

Cuba: 1953-1959

The cases included in the work, dating from 1933 to 1962, were selected to cover a wide range of outcomes, geographical regions, historical periods, and motivations. The asterisks represent more than one case study per state. The structure of the study follows that of *Casebook on Insurgency and Revolutionary Warfare Volume II*.

Human Factors Considerations of Undergrounds in Insurgencies

This study contains vignettes about many different insurgent organizations. Those with more extensive data are identified in the following table, and the chapter(s) in which the analysis pertaining to the organization is located is included in parentheses.

SOCAFRICA

Egypt: Egyptian Islamic Jihad (Chapters 4, 5, 11)

Egypt 2011 (Chapter 8)

South Africa (Chapter 6)

Algeria: National Liberation Front (Chapter 11)

SOCCENT	Al Qaeda (Chapters 3, 4, 7, 9, 11) Irgun (Chapter 8) Hamas (Chapter 8) Hizbollah (Chapter 8)
SOCEUR	Northern Ireland: Irish Republican Army (Chapters 3, 5, 6, 8, 11) Poland: Solidarity (Chapter 10) Ukraine: Orange Revolution (Chapter 10)
SOCPAC	Malaysia: Malayan Communist Party (Chapters 3, 11) Vietnam: Viet Minh (Chapters 5, 6) Viet Cong (Chapters 9, 11) Philippines: Philippino Hukbalahap (Chapters 5, 11) Sri Lanka: Tamil Tiger (Chapter 6, 8, 11)
SOCSOUTH	Venezuela: Fuerzas Armadas de Liberacion National (Chapter 3) Mexico: Zapatista National Liberation Army (Chapter 8) Colombia: 1948 (Chapter 10) Peru: Shining Path (Chapter 11)

This volume explores a wide variety of human factors that influence organizational and individual behaviors in conflict processes with an emphasis on resistance movements. It examines the impact of overarching structural factors, such as political or economic exclusions, on the emergence of conflict. Later chapters discuss the characteristics of resistance movements that influence the development, behaviors, and strategic and tactical decisions of resistance movements. Researchers also discuss the psychological risk factors associated with individual decisions to join, stay, and exit resistance movements as well as an extended treatment of leadership, including specific cognitive, psychodynamic, and personological approaches to assessing leadership.

Undergrounds in Insurgent, Revolutionary, and Resistance Warfare

This study provides an update to the original work of the same title, first published in 1963. This update seeks to account for the impact of major events, such as the Cold War and its end, on the underground components of insurgencies. While the underground was formerly illegal and clandestine, this 2013 volume suggests that over time, undergrounds operate in both illegal and legal domains. Furthermore, the boundaries between their clandestine and overt activities are blurred. Using the Tier I and Tier II case studies developed as part of ARIS, this study examines the anatomy of undergrounds, to include leadership and organization, recruiting, intelligence, finance, logistics, training, communications, security, shadow government, and subversion and sabotage.

2014

Case Studies in Insurgency and Revolutionary Warfare: Sri Lanka (1976-2009)

This study examines the evolution of insurgent activities in Sri Lanka over three decades. The two insurgent movements explored are the People's Liberation Front (JVP), a communist movement that employed terror tactics, and the Liberation Tigers of Tamil Eelam (LTTE), a nationalist movement representing the Tamil ethnic group. These case studies seek to elucidate the ways that these groups adapted to environmental and contextual challenges which include the physical environment, historical context, socioeconomic conditions, and government and politics. The structure and dynamics of each movement are also considered and include leadership and organization, ideology, legitimacy, motivation, the role of external actors, finances, etc. The historical tracing of the evolution of these movements will help the soldier to understand why and how both groups were ultimately defeated militarily without achieving their goals. While these movements were defeated, they exerted influence in other ways. After the publication of this study, this influence evolved into an electoral strategy where the JVP went on to found the United People's Freedom Alliance that performed well in the 2014 elections.

Case Studies in Insurgency and Revolutionary Warfare: Colombia (1964-2009)

This publication details case studies of insurgent groups operating in Colombia from 1964 to 2009. While there are numerous insurgent groups that could be considered during this period, this study focuses on the Revolutionary Armed Forces of Colombia (FARC), National Liberation Army (ELN), 19th of April Movement (M-19), and United Self-Defenders of Colombia (AUC). This work elucidates the ways that these groups adapted to environmental and contextual challenges, which include the physical environment, historical context, socioeconomic conditions, and government and politics. The structure and dynamics of each movement are also considered and include leadership and organization, ideology, legitimacy, motivation, the role of external actors, finances, etc. The historical tracing of the evolution of these movements will help the soldier to understand the roots of violence in La Violencia (Colombia's mid-century civil war), the government's pursuit of military and political solutions, and the government-movement interactions that yielded the current state of affairs in Colombia.

Legal Implications of the Status of Persons in Resistance

Determining the legal status of US military members conducting foreign internal defense, counterinsurgency, and UW is difficult. This is in large part because the legal status of US military members can change throughout the course of a single effort based on the legal status and recognition of the resistance movement involved. Grounding their work in an analysis of international and domestic law, the authors of this study present a novel solution to the difficulty of determining the legal status of US military members; they create a continuum of resistance that soldiers can use to determine their legal status at any point in time during deployment. This continuum is based on the resistance movement's activities and includes, from nonviolent to violent activities, nonviolent legal (the employment of legal processes to gain political advantage), nonviolent illegal (participating in illegal political acts), rebellion (violent), insurgency (violent), and belligerency (violent). This study reviews existing law, presents the continuum of resistance framework, and addresses US domestic legal constraints and current issues, such as US action in Syria.

2015

"Little Green Men": A Primer on Modern Russian Unconventional Warfare, Ukraine (2013-2014)

This case study of Russian activities in Ukraine was written to elucidate the conflict at that time. Russian involvement in Ukraine is a paradigmatic example of the Russian approach to UW called the Gerasimov model. This model is named for General Gerasimov, chief of general staff of the Russian Federation, who in writings described Russia's approach to modern warfare, the hallmarks of which are undeclared rather than declared wars; hybrid operations that combine military and non-military activities; the employment of smaller precision forces; the primacy of intelligence and domination of the information space through cyberwarfare, propaganda, and deception; and the use of proxies instead of uniformed Russian troops.

This report traces the history and evolution of this uniquely Russian approach to modern UW that started in the 1990s and occurred in many theaters, including Lithuania, Transnistria, Chechnya, and Georgia. After establishing a conceptual and historical basis for the Russian approach in previous theaters and in the Gerasimov doctrine, the report provides a detailed treatment of Russian UW in Ukraine. This treatment includes the actors involved, their motivations, the Russian order of battle, and Russian tactics, techniques, and procedures.

Case Studies in Insurgency and Revolutionary Warfare: Palestine Series Volume I – The Zionist Insurgency (1890-1950)

This case study explores the development of Zionism from its origins in the late 1800s through the founding of Israel in the late 1940s. In this study, the authors treat the Zionist movement as an insurgency aimed at overturning Ottoman, British, and Arab governance of Palestine and replacing it with a new form of governance—the Jewish state of Israel. This exploration addresses the contexts, catalysts, the structure, and dynamics of the insurgency and the countermeasures governments used to address it. Ultimately, the authors attribute the movement's success in founding a Jewish state of Israel to centripetal forces that unified the Zionist movement. These forces include historical anti-semitism, antipathy of neighboring Arab states, revival of the Hebrew language, strong economic and financial organization and investment that undergirded Jewish settlement in Palestine, and the Tanakh (Hebrew Bible). These centripetal forces outnumbered centrifugal forces that sought to pull the movement apart and include the Jewish diaspora's cultural, geographic, and linguistic diversity, varying conceptions of Zionist ideology, and different attitudes toward great power countries such as England.

Special Topics in Irregular Warfare: Understanding Resistance

This publication is an edited volume comprised of ten articles that address wide-ranging topics, all of which are intended to help the soldier better understand resistance and frame future research and analysis. Each article serves as a standalone piece for the ARSOF soldier to reference.

The articles cover topics such as organizational growth constructs of the resistance movement, the public component of a resistance movement, the interest, identification, indoctrination, and mobilization phases of recruitment, the threshold of violence to balance power to achieve a resistance movement's ends, information technology and threat finance employed by a resistance movement, methods of achieving legitimacy for a resistance movement, factors involved in a post-transition to governance, and external support leveraged for a resistance movement.

2016

Unconventional Warfare Pocket Guide Volume 1.0

This volume reviews UW theory, practices, and tactics, techniques, and procedures. Discussion is structured around the seven phases of UW: preparation, initial contact, infiltration, organization, buildup, employment, and transition. Each phase details a set of core activities for the soldier. The concluding section provides a guide to further resources on key UW topics, including a list of recommended sources on law and policy, doctrine, and academic research.

The UW pocket guide is designed to be a quick reference guide for the ARSOF soldier to identify terms, definitions, joint publication references, and doctrine.

Unconventional Warfare Case Study: The Relationship Between Iran and Lebanese Hizbollah

This case study explores the sponsor-proxy relationship between Iran and Hizbollah. This country and movement provide a paradigmatic example of the sponsor-proxy relationship. The authors explore Iran's motivations in sponsoring Hizbollah and the mechanisms through which it supports the movement. Hizbollah's organizational structure, methods of warfare, and political activities are also considered. The authors provide a detailed treatment of how Iran's relationship with Hizbollah was instrumental in checking Israeli expansion into Lebanon. Ultimately, the authors observe that Iran's sponsorship of Hizbollah successfully helped Iran meet its goals, not least of which was to hamper Israel's activities. The ideological affinity between Iran and Hizbollah helped to ensure a strong relationship, but challenges stemming from Iran delegating tasks to Hizbollah (principal-agent issues) still persist.

Unconventional Warfare Study Research and Writing Guide

The purpose of this research and writing guide is to assist JHU/APL and Special Forces personnel to create UW case studies. By following the approach outlined in this guide, the researcher and author can create a case study that complements and contributes to the work in the ARIS portfolio. The hallmark of the UW case study is a focus on the motivations, methods, and activities of the resistance movement's external supporter, whether a state or non-state actor. The guide recommends that any new case study comprises six sections: 1) an introduction that describes the research questions, the methodology, and the findings; 2) an overview of the external supporters discussed in the case study; 3) a discussion of the historical and socioeconomic factors that shape the environment in which the resistance operates; 4) an overview of the resistance movement, to include its structure and activities; 5) a description of the nature and effect of the external supporters' activities; and 6) a conclusion.

2018

Narratives and Competing Messages

This study defines narratives and identifies ways in which they can help soldiers understand and influence political and social behavior. Narratives are essentially about how individuals and groups experience the world and communicate that experience to others. As such, narratives are routinely used by insurgents, resistance movements, and counterinsurgents to mobilize people to support their causes. This publication reviews existing work on narratives and gives soldiers the tools and frameworks they need to develop persuasive messages, such as vertical integration (which works to connect personal and local narratives with narratives used by the society at large), myth-symbols, and imagery, among others. While this work can help to build a starting point for a soldier's strategic communication, deep knowledge of the local people and culture s/he is trying to influence is essential to effectively crafting messages. This work encourages soldiers to view narrative as another tool in their kit.

Understanding States of Resistance 🛡️

This volume reviews existing literature on phasing models depicting the life cycle of insurgencies and resistance movements. Building on existing material, the work constructs a continuum of phases of a resistance movement that emphasizes nonlinear or nonsequential shifts along the continuum. Existing ARIS case studies are coded according to its transition paths and examined to identify the organization, infrastructure, resources, leadership, and other factors that facilitate transition of resistance movements forward or backward along the phasing continuum.

There are fourteen detailed vignettes that align the movement to the state (phase) of the referenced resistance.

SOCCENT		SOCEUR	
Iran Revolution	Chechnya: Chechen Revolution		
Lebanon: Hizbollah	Northern Ireland: IRA		
Palestine: PLO	Poland: Solidarity		
Taliban	Ukraine: Orange Revolution		
SOCPAC		**SOCSOUTH**	
Burma: KLNA	Colombia: FARC		
Sri Lanka: LTTE	El Salvador: FMLN		
Vietnam: Viet Cong	Peru: Shining Path		

The Science of Resistance 🛡️

This study seeks to provide the soldier with an overview of answers to some challenging questions: Why does resistance occur in some places and not others? Why do some people join a resistance and others do not? How do resistance movements choose the tactics they employ? Why do conflicts end? These challenging questions have been rigorously addressed, over decades, in the fields of political science and sociology.

The intent of this study is to provide the introduction to political science and sociology course to answer and help frame these questions for the soldier in an operationally relevant way. To answer the question about why resistance occurs in some places and not others, the authors explore explanations based on grievances, geography, quality of governance, and political and international factors. To shed light on why some people are motivated to join a resistance while others are not, the authors explore the collective action challenge and identify risk factors for joining a resistance movement. To understand why resistance movements choose certain tactics, the authors discuss the choice to transition from nonviolent to violent tactics and the impact to organizational structure and resourcing. Finally, to illuminate why conflicts end, the authors present research on the various types of civil war termination (negotiated settlement, cease fire, etc.) and the evolution of social movements. It is important to note that the social science disciplines have not definitively answered these questions. However, there are a number of complementary and competing theories that shed light on the answers and better inform the soldier's actions.

Conceptual Typology of Resistance

This study creates a conceptual typology of resistance. A typology is useful to the soldier because it breaks a complex phenomenon into its constituent parts. This makes the phenomenon easier to understand, successfully interact with, and analyze. First, the authors identify the fundamental attributes of a resistance as: actors, causes that motivate the actors, environment, organization of the resistance movement, and actions of the resistance movement. Next, the authors divide each of these attributes into categories. For the attribute of the environment, for example, the authors identify and consider the environment the state creates and the environment created by counterinsurgency efforts. The authors developed this conceptual typology by following social science methodologies and further informed the process with the insights of social science subject matter experts and the USASOC community during multiple collaborative analysis events.

2019–2020

Threshold of Violence

This study explores the challenge posed by a resistance movement's use of violence, namely, the acceptable threshold of violence employed? In using violence, a resistance movement needs to ensure that the amount does not alienate supporters and conveys strength to the government it resists. The challenge of how much violence to use and when is captured graphically in the literature's equivalent response model, which seeks to identify an acceptable band of violence, with a lower and upper threshold. This study accomplishes three aims. First, it uses rational choice theory to explain how resistance movements make decisions about the use of violence. Second, it discusses some of the operational benefits of the use of violence, such as how its use can help a movement accomplish its aims. Third, the study explores tactics that resistance movements often use to raise the upper threshold of violence. Those tactics include provoking an indiscriminate government counterresponse to increase the tolerance of the resistance movement's employment of violence, providing public goods and social welfare to the population to gain popular support, and using narratives to help shape community standards regarding legitimate acts of violence.

Case Studies in Insurgency and Revolutionary Warfare: Palestine Series Volume II – The Palestinian Arab Insurgency (1890-2010)

This volume is a comprehensive review of the long-running political conflict processes in Palestine from 1890 to 2010 with a focus on successive Palestinian resistance movements. The initial sections examine the impact of structural factors—physical environment, historical context, socioeconomic conditions, and government politics—on political conflict processes in the region. Later sections address twelve crucial periods, including the origins of the Arab resistance in 1890-1914, Israel's Operation CAST LEAD of 2008-2009 in the Gaza Strip, and the aftermath of that operation. The resistance movements critical in each period are also discussed, analyzing the movement's origins, individual and organizational behavior in conflict processes, and the impact of dynamic interactions between the resistance movement and state security forces.

Case Studies in Insurgency and Revolutionary Warfare: The Patriot Insurgency

This volume is an overview of the political conflict processes of the American colonies in 1763-1789. The initial sections examine the impact of structural factors—physical environment, historical context, socioeconomic conditions, and government politics—on political conflict processes. Later sections address the Patriot resistance movement, analyzing the movement's origins, individual and organizational behavior in conflict processes, and the impact of dynamic interactions between the resistance movement and British security forces. The ARSOF soldier can leverage the role of organizational structure and external support, in particular, to inform his/her understanding of resistance networks.

Unconventional Warfare Case Study: The Rhodesian Insurgency and the Role of External Support (1961-1979)

This volume analyzes Chinese and Russian external support to the Zimbabwe African People's Union and the Zimbabwe African National Union in Rhodesia. It examines the social, historical, political, and economic context that shaped the states' UW. Initial sections also addresses the organizational characteristics, strategies, tactics, and operations of the resistance movements. The final chapter compares Russia's and China's UW campaigns in Rhodesia, including the actors, methods, and forms of support used by each state.

The Day After Overthrow: How the Behavior of the State and the Resistance Shapes Post-Overthrow Outcomes

This study discusses the characteristics of a resistance movement and the state it tries to overthrow that impact the nature of post-conflict governance. The report assesses this impact by investigating cases of resistance movements overthrowing states in the post World War II time period. The report discusses and offers a novel approach to conceptualizing state overthrow, presenting the magnitude of overthrow scale. The study also reviews assessed variables and how they were operationalized. A statistical analysis was performed on cases that fall on two rungs on the magnitude of overthrow scale—forced removal of the head of state and forced constitutional reform. This statistical analysis yielded correlative findings. Further study of the magnitude of the overthrow scale is suggested to reveal causal findings and provide findings relevant to all types of overthrow.

SOCAFRICA	SOCCENT	SOCEUR	SOCPAC	SOCSOUTH
Benin*	Kyrgyzstan	Romania	Cambodia	Argentina
Chad	Pakistan	Czechoslovakia	Fiji*	Brazil
Cote d'Ivoire		Germany East	Philippines	Haiti
Guinea-Bissau*		Greece	Thailand*	Peru*
Uganda*		Hungary	South Korea	Guatemala
Burkina Faso		Spain		
Burundi				
Congo Kinshasa				
Ghana				
Madagascar				
Nigeria*				
Senegal				
Sudan				

Forced Removal of Head of Government

Forced Constitutional Reform

Both

This discussion of findings distinguishes factors that ARSOF soldiers should be able to influence from those that are structural and difficult to influence. The study identifies thirty-nine cases of polity transition from the Polity IV dataset, created and managed by the Center for Systemic Peace. Those marked with an asterisk are states with more than one instance of a polity transition.

ARIS Codebook

The Codebook is a reference and analytic tool to aid users in identifying available social science data related to resistance as well as existing gaps in data. Researchers identified thirty-seven datasets related to resistance and accompanying codebooks for analysis. The thirteen hundred variables are mapped into a matrix that cross references existing resistance data with the ARIS *Conceptual Typology of Resistance*. The Codebook includes units of analysis and decryptions of the variables for each dataset. Statistical analysis of the matrix identifies areas of data saturation and data gaps.

Nonconventional Assisted Recovery (NAR)

US military personnel often deploy to areas with a risk of separation from their units and capture by the enemy. Many of these occurrences were popularized in books and movies such as *Bat 21* that depicts an Air Force navigator on an electronic warfare aircraft is shot down behind enemy lines in Vietnam and *Lone Survivor* (the story of Navy Seal Marcus Luttrell who becomes isolated after a mission against the Taliban goes awry). It is often possible to rescue isolated personnel using *conventional* personnel recovery capabilities, such as combat search and rescue (US Air Force) or tactical recovery of aircraft personnel (US Marine Corps), fielded by the Services. However, in enemy-controlled areas, the risk to conventional recovery forces may be too great to chance a rescue attempt, and another solution is needed. Non-conventional assisted recovery (NAR) relies on indigenous personnel, recruited by US Special Forces or the intelligence community, to contact and support US isolated personnel behind enemy lines and assist with their eventual recovery by friendly forces. Building a NAR capability in a foreign country that remains viable, perhaps over many years, until it is needed is a complex challenge. This study reviews historical examples of NAR and current practices to build NAR networks to derive enduring principles for the successful development, maintenance, and execution of NAR capabilities.

Resistance in the Cyber Domain

In recent years, it has become clear that the next major war is going to be a war where the acquisition, denial, and employment of information will be an important weapon. The growth of the cyber domain, with its concomitant abilities to vector huge quantities of information (sometimes accurate, sometimes not) to billions of Internet users worldwide at trivial cost exacerbated the importance of information operations across the spectrum of communications and conflict. In a world that witnessed the rise of insurgencies across the globe, the Internet is a weapon used not only by hackers seeking to empty the bank accounts of unwitting victims but also by governments dedicated to the defeat, or overthrow, of their enemies.

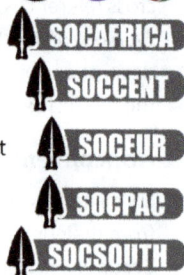

This work explores some recent resistance movements and highlights their methods, successes, and failures, as well as the efforts of their adversaries. The purpose of this study is to offer soldiers an overview of a new realm of warfare, particularly as it applies to resistance movements and information operations. Topics include cyber resistance, cyber operations, cyber security, narratives and social media, cyber physical systems, cyber underground, movement attribution, and legal issues. Key takeaways from each chapter provide soldiers with guiding principles to inform their actions in the cyber realm. These key takeaways are crucial for the development of successful cyber operations and the development of appropriate security protocols. The final chapter contains fictional examples and real case study scenarios that integrate the key takeaways into an applied discussion.

ARIS Resistance Manual

This draft manual seeks to provide the introduction to resistance studies for the ARSOF soldier. The soldier follows the journey of an ARSOF soldier deployed to the fictional country of Estatu where he meets a rising resistance leader named Arturo Bolonieves. As Bolonieves mobilizes the Sarca resistance, the ARSOF soldier surveys the science of resistance to advise him on the actors, causes, environment, organization, and actions of the resistance movement. Drawing from the entire body of ARIS studies and infusing cutting-edge academic research, the Manual provides the fundamentals of resistance with application-based discussion questions for the soldier to consider.

PART III: CONCLUSION

In 2009, the ARIS program was created to provide the ARSOF soldier
with foundational knowledge concerning the object of his/her
profession—resistance. Much like surgeons guided by the biological
sciences in their chosen craft, ARSOF soldiers must rigorously
study, practice, and apply the body of scientific knowledge to their
profession. To effectively support and/or counter a resistance, ARSOF
soldiers benefit from understanding the advancements in their
chosen field of study.

By placing ARIS within the broader context of academic literature,
one can easily identify the importance of significant topics,
publications, methodologies, and researchers in the science of
resistance. Employing bibliometric analysis of political science,
anthropology, sociology, psychology, economics, history, and law,
this pocket guide demonstrated the prolific research in resistance,
conflicts, violence, strife, civil disobedience, ethnic conflicts,
insurrections, rebellions, and coup d'états. The use of case study
methodologies on the American Civil War, Spain, Uganda, and
other various African countries established the components of
the academic pursuit of resistance studies. Scholars published
significant, paradigm-shifting analysis in the fields of conflict drivers,
rebel actors, and regime changes. These studies can frequently
be found in such significant journals as the *Journal of Peace
Research*, *Democratization*, *Journal of East African Studies*, *Geoforum*,
Journal of Economic History, and *Latin American Studies*. Finally, the
paradigm-defining researchers, such as Mancur Olson, Charles Tilly,
Theda Skocpol, Ted Robert Gurr, and Samuel Huntington, were
identified along a chronological timeline of their publications.
Creating a visual depiction of these significant pieces of resistance

studies demonstrated the interconnectedness of a study to other subsequent important studies.

The ARIS publications complement these advances in resistance studies by providing rigorous research that can be used by and useful to ARSOF soldiers in their deployments. Mr. Paul Tompkins' experience as a special operations force operator, planner, and now director of the Sensitive Activities Division demonstrates the importance of leveraging the rigorous examination of resistance studies to inform and execute field activities on deployment. The evolution of ARIS from historical case studies to topical studies and now scientific methodological studies strives to be repeatable, reliable, explainable, and traceable. The future of ARIS research must continue to be operationally relevant and rigorously applied to meet the mission requirements of the ARSOF soldier.

APPENDIX

This appendix details the color coding used in Figures 3 and 4.

Causes of Conflicts, Theory, Prevalence	5.6%
English Civil War	4.8%
Democratization, Regime Change, Use of force by democracies	4.3%
Russia, Soviet, Post Soviet, and interactions with Afghanistan	4.3%
Ethnic conflict	4.3%
Civil wars and Rebels	3.9%
MENA (generally)	3.8%
Methods, Models, Mobilization	3.3%
Holocaust and WWII Resistance	3.1%
Race and US Civil Rights	2.9%

Religion	2.8%
Slavery, Emancipation, War Commemoration	2.7%
Nineteenth Century Wars	2.5%
Women and Gender	2.5%
Elections	2.5%
Secession, Legal Aspects	2.4%
Spanish Civil War	2.4%
Natural Resources and Conflict	2.4%
Refugees, Forced Migration, Camps	2.3%
Peacekeeping and Third Party Intervention	2.2%
Modeling and operations research methods	2.1%

British in India	2.1%	Civil Disobedience	1.5%
Terrorism and Counterterrorism	2.0%	Al Qaeda and Osama bin Laden	1.4%
Media (social and traditional)	2.0%	Sri Lanka	1.4%
Insurgencies	1.9%	Post-Cold War	1.3%
Israeli Palestinian Conflict	1.9%	Latin America and the Caribbean	1.2%
Ireland	1.8%	Activism, Collective Action, Beginnings	1.2%
Land rights and Natural Resources	1.8%	Sierra Leone	0.92%
China	1.7%	Korea	0.82%
South Africa	1.7%	Sudan, South Sudan	0.81%
Peace, Demobilization, Resolution	1.7%	El Salvador	0.68%
Human rights and transitional justice	1.6%		
Constitutions and Courts	1.6%		

NOTES

[1]   Michael Collins, *A Path to Freedom* (Boulder, CO: Roberts Rinehart Publishers, 1996).

[2]   US Joint Chiefs of Staff, "Special Operations," Joint Publication 3-05, April 18, 2011, https://fas.org/irp/doddir/dod/jp3-05.pdf.

[3]   Jonathon B. Cosgrove and Erin N. Hahn, *Conceptual Typology of Resistance* (Ft. Bragg, NC: US Special Operations Command, 2018).

[4]   Ted Robert Gurr, *Why Men Rebel* (Princeton, NJ: Princeton University Press, 1970).

[5]   Ted Robert Gurr, "Why Men Rebel Redux: How Valid are its Arguments 40 Years On?," *E-International Relations*, November 17, 2011, accessed July 9, 2018, http://www.e-ir.info/2011/11/17/why-men-rebel-redux-how-valid-are-its-arguments-40-years-on/.

[6]   Samuel P. Huntington, *The Clash of Civilizations and the Remaking of World Order* (New York: Simon & Schuster, 1996).

[7]   James D. Fearon, K. Kasara, and D. D. Laitin, "Ethnic Minority Rule and Civil War Onset," *American Political Science Review* 101, no. 1 (February 2007), http://dx.doi.org/10.1017/S0003055407070219; James D. Fearon and David D. Laitin, "Ethnicity, Insurgency, and Civil War," *American Political Science Review* 97, no. 1 (2003), http://www.worldcat.org/oclc/365081096; J. D. Fearon and D. D. Laitin, "Explaining Interethnic Cooperation," *American Political Science Review* 90, no. 4 (December 1996), http://dx.doi.org/10.2307/2945838.

[8]   Haivard Hegre, et al., "Toward a Democratic Civil Peace? Democracy, Political Change, and Civil War, 1816-1992," *American Political Science Review* 95, no. 1 (2001); H. Hegre, et al., "Predicting Armed Conflict, 2010-2050," *International Studies Quarterly* 57, no. 2 (June 2013), http://dx.doi.org/10.1111/isqu.12007.

[9]   Václav Havel, *The Power of Powerlessness: Citizens Against the State in Central Eastern Europe* (New York: Routledge Press, 1985).

BIBLIOGRAPHY

Agan, Summer D., et al. *Science of Resistance*. Ft. Bragg, NC: US Special Operations Command, 2018.

Baum, Matthew A. "Soft News and Political Knowledge: Evidence of Absence or Absence of Evidence?" *Political Communication* 20, no. 2 (2003): 173-190.

Blattman, Christopher and J. Annan. "The Consequences of Child Soldiering." *Review of Economics and Statistics* 92, no. 4 (2010): 882-898.

Blattman, Christopher. "From Violence to Voting: War and Political Participation in Uganda." *American Political Science Review* 103, no. 2 (2009): 231-247.

Bos, Nathan. *Human Factors Considerations of Undergrounds in Insurgencies*. Second Edition. Ft. Bragg, NC: US Special Operations Command, 2013.

Buhaug, Halvard, S. Gates, and P. Lujala. "Geography, Rebel Capability, and the Duration of Civil Conflict." *Journal of Conflict Resolution* 53, no. 4 (2009): 544-569.

Cohen, Dara K. and R. Nordås. "Sexual Violence in Armed Conflict: Introducing the SVAC Dataset, 1989–2009." *Journal of Peace Research* 51, no. 3 (2014): 418-428.

Collier, Paul and Anke Hoeffler. "On Economic Causes of Civil War." *Oxford Economic Papers* 50, no. 4 (1998): 563-573.

Collier, Paul. *The Bottom Billion: Why the Poorest Countries are Failing and What Can Be Done About It*. Oxford: Oxford University Press, 2007.

Collins, Michael. *A Path to Freedom*. Boulder, CO: Roberts Rinehart Publishers, 1996.

Cosgrove, Jonathon and Erin N. Hahn. *Conceptual Typology of Resistance*. Ft. Bragg, NC: US Special Operations Command, 2018.

Crossett, Chuck. *Casebook on Insurgency and Revolutionary Warfare Volume II: 1962-2009*. Ft. Bragg, NC: US Special Operations Command, 2012.

Ellison, Melissa, et al. *The Day After Overthrow: How the Behavior of the State and the Resistance Shapes Post-Overthrow Outcomes*. Ft. Bragg, NC: US Special Operations Command, 2019.

Fearon, James D. and David D. Laitin. "Ethnicity, Insurgency, and Civil War." *American Political Science Review* 97, no. 1 (2003). http://www.worldcat.org/oclc/365081096.

Fearon, James D. and David D. Laitin. "Explaining Interethnic Cooperation." *American Political Science Review* 90, no. 4 (1996). http://dx.doi.org/10.2307/2945838.

Fearon, James D., K. Kasara, and David D. Laitin, "Ethnic Minority Rule and Civil War Onset." *American Political Science Review* 101, no. 1 (2007). http://dx.doi.org/10.1017/S0003055407070219.

Fjelde, Hanne and D. Nilsson. "Rebels Against Rebels: Explaining Violence Between Rebel Groups." *Journal of Conflict Resolution* 56, no. 4 (2012): 604-628.

Fjelde, Hanne and Lisa Hultman. "Weakening the Enemy: A Disaggregated Study of Violence Against Civilians in Africa." *Journal of Conflict Resolution* 58, no. 7 (2014): 1230-1257.

Foner, Eric. *Reconstruction: America's Unfinished Revolution, 1863-1877*. 1st Perennial Classics ed. New York: Perennial Classics, 2002.

Gleditsch, Kristian S. and M. Rivera. "The Diffusion of Nonviolent Campaigns." *Journal of Conflict Resolution* 61, no. 5 (2017): 1120–1145.

Gleditsch, Nils P. "Whither the Weather? Climate Change and Conflict." *Journal of Peace Research* 49, no. 1 (2012): 3-9.

Gleditsch, Nils P., P. Wallensteen, M. Eriksson, M. Sollenberg, and H. Strand. "Armed Conflict 1946-2001: A New Dataset." *Journal of Peace Research* 39, no. 5 (2002): 615-637.

Grdovic, Mark, et al. *Unconventional Warfare Case Study: The Rhodesian Insurgency and the Role of External Support: 1961-1979.* Ft. Bragg, NC: US Special Operations Command, 2019.

Gurr, Ted Robert. "Why Men Rebel Redux: How Valid are its Arguments 40 Years On?" *E-International Relations* (November 17, 2011). Accessed July 9, 2018. http://www.e-ir.info/2011/11/17/why-men-rebel-redux-how-valid-are-its-arguments-40-years-on/.

Gurr, Ted Robert. *Why Men Rebel.* Princeton, NJ: Princeton University Press, 1970.

Hahn, Erin N. and W. Sam Lauber. *Legal Implications of the Status of Persons in Resistance.* Ft. Bragg, NC: US Special Operations Command, 2014.

Hahn, Erin N., et al. *Special Topics in Irregular Warfare: Understanding Resistance.* Ft. Bragg, NC: US Special Operations Command, 2015.

Havel, Vaclav. *The Power of Powerlessness: Citizens Against the State in Central Eastern Europe.* New York: Routledge Press, 1985.

Hegre, Haivard, et al. "Predicting Armed Conflict, 2010-2050." *International Studies Quarterly* 57, no. 2 (2013). http://dx.doi.org/10.1111/isqu.12007.

Hegre, Haivard, et al. "Toward a Democratic Civil Peace? Democracy, Political Change, and Civil War, 1816-1992." *American Political Science Review* 95, no. 1 (2001): 33-48.

Hoeffler, Anke. "Out of the Frying Pan into the Fire? Migration from Fragile States to Fragile States." *OECD Development Co-operation Working Papers* No. 9. OECD Publishing. 2013.

Humphreys, Macartan and J. M. Weinstein. "Demobilization and Reintegration." *Journal of Conflict Resolution* 51, no. 4 (2007): 531-567.

Humphreys, Macartan and J. M. Weinstein. "Handling and Manhandling Civilians in Civil War." *American Political Science Review* 100, no. 3 (2006): 429-447.

Humphreys, Macartan and J. M. Weinstein. "Who Fights? The Determinants of Participation in Civil War." *American Journal of Political Science* 52, no. 2 (2018): 436-455.

Huntington, Samuel P. *The Clash of Civilizations and the Remaking of World Order.* New York: Simon & Schuster, 1996.

Kalyvas, Stathis N. *The Logic of Violence in Civil War.* New York: Cambridge University Press, 2006.

Kocher, Matthew. "State Capacity as a Conceptual Variable." *Yale Journal of International Affairs* 5 (2010): 137-145.

Lauber, W. Sam, et al. "Resistance Manual." ARIS Publication, 2019.

Lauber, W. Sam, et al. *Understanding States of Resistance.* Ft. Bragg, NC: US Special Operations Command, 2018.

Leonhard, Robert and Stephen P. Phillips. *Case Studies in Insurgency and Revolutionary Warfare – Palestine Series Volume II – The Palestinian Arab Insurgency (1890-2010).* Ft. Bragg, NC: US Special Operations Command, 2019.

Leonhard, Robert, et al. *Case Studies in Insurgency and Revolutionary Warfare – Palestine Series Volume I – The Zionist Insurgency (1890-1950).* Ft. Bragg, NC: US Special Operations Command, 2015.

Leonhard, Robert, Summer D. Agan, and Stephen P. Phillips. *Case Studies in Insurgency and Revolutionary Warfare – The Patriot Insurgency (1763-1789)*. Ft. Bragg, NC: US Special Operations Command, 2019.

Leonhard, Robert. "'Little Green Men:' A Primer on Modern Russian Unconventional Warfare, Ukraine 2013-2014." ARIS Publication, 2015.

Leonhard, Robert. *Undergrounds in Insurgent, Revolutionary, and Resistance Warfare*. Second Edition. Ft. Bragg, NC: US Special Operations Command, 2013.

Lyall, Jason and A. Dafoe. "From Cell Phones to Conflict? Reflections on the Emerging ICT-Political Conflict Research Agenda." *Journal of Peace Research* 52, no. 3 (2015): 401-413.

Lyall, Jason. "Pocket Protests: Rhetorical Coercion and the Micropolitics of Collective Action in Semiauthoritarian Regimes." *World Politics* 58 (2006): 378-412.

Miers, Jeff, et al. "Non-Conventional Assisted Recovery – NAR." ARIS Publication, 2019.

Newton, Summer D. "Irregular Warfare Annotated Bibliography." ARIS Publication. June 2, 2011.

Newton, Summer D. et al. *Codebook (2016-2017)*. Ft. Bragg, NC: US Special Operations Command, 2019.

Newton, Summer D., et al. *Case Studies in Insurgency and Revolutionary Warfare – Colombia (1964-2009)*. Ft. Bragg, NC: US Special Operations Command, 2014.

Newton, Summer D., et al. *Narratives and Competing Messages*. Ft. Bragg, NC: US Special Operations Command, 2018.

Pinczuk, Guillermo and Chris Kurowski. *Threshold of Violence*. Ft. Bragg, NC: US Special Operations Command, 2019.

Pinczuk, Guillermo and Theodore Plettner. *Unconventional Warfare Case Study: The Relationship Between Iran and Lebanese Hizbollah*. Ft. Bragg, NC: US Special Operations Command, 2017.

Pinczuk, Guillermo, Mike Deane, and Jesse Kirkpatrick. *Case Studies in Insurgency and Revolutionary Warfare – Sri Lanka (1976-2009)*. Ft. Bragg, NC: US Special Operations Command, 2014.

Ryan, Kristen, et al. *Resistance and the Cyber Domain*. Ft. Bragg, NC: US Special Operations Command, 2019.

Sambanis, Nicholas and M. Shayo. "Social Identification and Ethnic Conflict." *American Political Science Review* 107, no. 2 (2013): 294-325.

Soderbom, Mans, Paul Collier, and Anke Hoeffler. "Post-Conflict Risks." *Journal of Peace Research* 45, no. 4 (2008): 461-478.

Toft, Monica D. "Ending Civil Wars: A Case for Rebel Victory?" *International Security* 34, no. 4 (2010): 7-36.

Toft, Monica D. "Population Shifts and Civil War: A Test of Power Transition Theory." *International Interactions* 33, no. 3 (2007): 243-269.

Toft, Monica D. "Territory and War." *Journal of Peace Research* 51, no. 2 (2014): 185-198.

Tonnon, Joseph, et al. "Unconventional Warfare Study Research and Writing Guide." ARIS Publication. 2017.

US Special Operations Command. "Special Operations." Joint Publication 3-05. April 18, 2011. https://fas.org/irp/doddir/dod/jp3-05.pdf.

US Special Operations Command. "Unconventional Warfare Pocket Guide." Volume 1. 2016.

Zhukov, Yuri M. and Monica D. Toft. "Islamists and Nationalists: Rebel Motivation and Counterinsurgency in Russia's North Caucasus." *American Political Science Review* 109, no. 2 (2015): 222-238.

www.ingramcontent.com/pod-product-compliance
Lightning Source LLC
Chambersburg PA
CBHW070030030426
42335CB00017B/2371